Contents

Are You Optimized? Common Signs of Brain Fatigue 7

Molding Your Mind: The Truth about Neuroplasticity 15

Brain Hacks: Proven Paths to Cognitive Enhancement, Memory Consolidation and Cerebral Dominance 18

Rewind for Wisdom -- 7 Practical Strategies to Boost the Aging Brain......33

Dark Psychology: Unleashing the Power of Manipulative Mind-Control...53

Master Mind:

Unleashing the Infinite Power of the Latent Brain

By C.K. Murray

Copyright © 2014

C.K. Murray
All Rights Reserved

Your brain is a *powerhouse*.

Every moment, every fleeting, seemingly insignificant moment, your brain is exploding. Impulses and chemicals and an endless array of messages fire across that bundle of grey matter, inspiring the million and one things that make you, *you*.

All of your actions, your behaviors, your thoughts and attitudes, all are tied to your flurried mind. When you choose coffee over water, fun over work, love over hate, this person over that person—every decision, whether life-changing or miniscule, is connected to the crazy nature of nature's most dynamic organ.

No matter who you are or what you do, your brain is amazing. Nerve impulses to and from the brain zip at the speed of a luxury sports car, exceeding 170 miles per hour. Your brain is buzzing with electricity, enough to turn on a light bulb, enough mental power to contain over 5 times the information stored in the Encyclopedia

Britannica. It is 80% water and 20% oxygen—60% white matter and 40% grey.

And filling that teeming, crazy tangle of neurons and possibility: oh, nothing impressive, just 150,*000* miles of blood vessels; tight, powerful, and pumping with fury.

Put simply, the brain is an incredible organ. And we should all take care of it.

But maybe you want more than to simply *care* for your powerful noggin. Maybe you want to go beyond, so far ahead of the curve, ahead of your peers, that your mental powerhouse reaches an all new level.

Tired of fatigue? Tired of feeling like your head is foggy and your thoughts sluggish? Wishing there was something you could do to really spruce up your thinking? A way to improve cognitive performance, boost productivity, enhance memory and optimize your brain for the long-haul? Do you *really* want to get that baby pulsating?

Well then stop fretting! It's time to master your mind…

Are You Optimized? Common Signs of Brain Fatigue

Before we get into innovative strategies for enhanced cognitive functioning, we need to take a very hard look at how we're doing. In other words, we need to understand how we can improve. For many of us, life is crazy and hectic. We have so much to do, that sometimes we feel like we're drowning in our own private rapids just to get it all done!

But what if all of this could become *significantly* easier? What if instead of going through turbulent waters with a paddleboat, you can tear through on a battleship—equipped with enough weaponry and defense systems to conquer even the toughest obstacles?

Well you can! And it all begins with knowing where you stand. Here are some of the most common signs of brain fatigue, a condition that can vary from acute episodes we all experience, to chronic debilitation that makes our lives unmanageable:

Memory Deficits

When it comes to using our brains, memory is incredibly important. There are numerous forms and functions for memory, everything from remembering how to tie your shoe, to knowing the route you take to get to work, and recalling the day you met that special someone. If you don't have your memory in working order, the more pressing things—like projects, meetings, and important calculations—all go out the window.

When you are suffering from brain fatigue, your memory is far from seamless. See, the human brain retains information through the process of acquisition, consolidation, and recall. Acquisition is when you are absorbing or studying the information, consolidation is when that information gets integrated into your brain and 'sticks,' and recall is, well, the ability to actually call up that information. During brain fatigue, we struggle with recall. We'll say something to the effect, "Oh, I know this, I know this!" or maybe we'll simply have a general feeling that we learned something at some point, but just can't seem to recall it at the moment. Brain fatigue will also cause problems with acquisition and consolidation. If we struggle with acquisition, information is more or less going in one ear and out the other. And if we struggle with consolidation, we might hear the information and think we understand it, but then we soon begin to struggle. We mix it up in our mind, we forget how it 'fits' with other things we know, and we end up scratching our heads and saying, "Wait… what was that again?"

Concentration and Processing

Obviously, this ties into memory.

When we struggle with concentration, everything seems to be going at breakneck speed. All the different stimuli rush by us as we try to keep pace. Different conversations are hard to follow. We may hear a noise there or a sound here, and simply trying to interpret and think about one stimulus will keep us distracted from another. We may become overburdened, finding multitasking almost impossible. Our reading will be slower and more tedious, changes in television like commercials and channels will be hard to follow, traffic will seem more intense, and virtually any situation involving large congregations of people will give us trouble. Completing tasks at work and taking part in social events will become much more difficult. Either we'll feel as if everything is going too fast, or we're feel like nothing is going at all, stuck inside our own heads with our untraceable thoughts.

Increased Stress & Emotional Sensitivity

Mental and physical stress can go hand-in-hand. With all the problems in our memory, concentration, and processing speed, we're bound to feel more stressed. We'll lose ourselves amid all our worries; we'll feel overwhelmed, even when the tasks were once manageable; we'll begin to experience physical maladies, like lethargy, fatigue, headaches, tension, shortness of breath, lightheadedness, sweating, elevated blood pressure, heart palpitations, and even more serious conditions. These symptoms may become so overpowering, that we'll forget how to cope. We'll lose sight of how to harness stress in a conducive manner. In the end, our minds and bodies will only continue to deplete. We'll be prone to mood swings; we'll become emotionally unpredictable to ourselves and to others. Irritability, anger, sadness, and even depression may result.

Poor Sleep

This one is a no-brainer result of all the other problems caused by brain fatigue. When your brain is running at sub-optimal levels, you're going to struggle. Although there are amazing ways to naturally improve sleep, many of us need to go further. People suffering from brain fatigue know what it's like to lack sleep. They either cannot get to sleep or they sleep too much. Some sleep for short periods or wake up constantly during the night. Severe brain fatigue may even lead to sleep disorders such as restless leg syndrome and narcoleptic behaviors (ie; nodding out at work, home, while driving, etc.).

Noise and Light Intolerability

Have you ever been really tired with your head cloudy, and noises and lights seeming too harsh and bright? Well, then you know quite well what this one's about. In some cases, sunglasses or earplugs are a temporary solution. Unfortunately, these strategies *will not* offer long-term solutions to the source of the issue.

Put simply, there are many symptoms and signs of brain fatigue. In most cases, these problems occur as a result of *unrestricted stimuli*. Unrestricted stimuli describes the invasion or overload of information that sabotages our brains both acutely and chronically. When your brain is forced to take in stimuli in an unprepared fashion, you will likely begin to deteriorate. Hectic and noisy environments can exacerbate this, as can difficult interactions with numerous people. When an individual begins to suffer from brain fatigue, even the simple things become tough. Reading, crowded public places, conversations, focusing, remembering, and dealing with unexpected events *all* become much more difficult if our brain is functioning at sub-optimal levels.

And when it comes to changing the game, the simple things simply won't do. Sure, you can take regular breaks, and you can take more rest, and you can work more steadily, take time to yourself, and try to be more productive—but what does all this *really* do?

If you want to go beyond simply treating brain fatigue, if you want to *optimize* your brain, and *master* your mind, now's the time.

Molding Your Mind: The Truth about Neuroplasticity

It's time to take charge of your brain. Unlocking the power of your mind is about more than simply thinker clearer. When our brains are optimized, they allow our bodies to function at a higher level as well. We end up spending less time on insignificant activities. Our reflexes become crisper, our thoughts and feelings are more positive, more powerful. We can think quicker and act quicker, significantly reducing the resources required for everyday tasks.

In a matter of time, what used to be difficult becomes automatic.

Once upon a time, the experts knew very little. In fact, up until just recently, most experts believed that our brains began to deteriorate from age and injury at around 25 to 30 years of life. In their minds, the growth and development of gray matter slowed, and the deathly spiral began. Nowadays, we know that this is simply not true.

Brain plasticity, also known as neuroplasticity or cortical remapping, refers to what we now know. Modern research reveals that the brain continues to create new neural pathways and alter preexisting ones, all in order to adapt, to create new memories, and promote new channels for learning, thinking and feeling in different ways.

In other words, the brain is more malleable than we ever thought. The more you do something, the more your brain reinforces that something by creating new neurons through a process known as neurogenesis. When you begin to do something new, your brain adapts to that too. Quite literally, the brain 'rewires' itself.

The Science of Plasticity

Again, plasticity begins in the brain. Plasticity means ability to change, and when it comes to neural changes, there are roughly 100 billion neurons susceptible to said change. And roughly 2,500 to 15,000 synapses per neuron.

Brain plasticity is an amazing thing. It is both environmental and genetic. It is never the same; it varies with age, with different changes at different times (puberty much?). Plasticity also incorporates brain cells aside from neurons, including glial and vascular cells, and can result as a point of learning, and as a result of brain damage. This explains why some people will actually learn to use different parts of their brains following severe brain trauma. As incredible as it is, functions once believed reserved for one segment or hemisphere of the brain will be relocated. This is what is called 'functional plasticity.' The brain's ability to actually change the physical structure of neural pathways is called 'structural plasticity.'

But enough talk about plasticity. Let's get down to actually changing those neurons!

Brain Hacks: Proven Paths to Cognitive Enhancement, Memory Consolidation and Cerebral Dominance

There are plenty of easy and incredible ways to improve your performance. If you're looking to think clearer and deeper, then look no further. The following strategies will allow you to quickly improve your ability to strengthen your cognitive functioning. They are <u>health hacks</u> specifically tailored to cognitive enhancement.

Let's get to 'em!

Vanilla Power

This tasty plant extract has been used as an antioxidant and brain booster for years. In a nutshell, vanilla contains vanilloids which reduce inflammation and improve mental performance. Vanilla is also effective at reducing stomach discomfort and hunger pangs.

The reason that vanilla improves cognitive performance is because it reduces inflammation. Inflammation occurs during periods of stress and can lead to the release of cytokines in the brain—basically causing that feeling of cloudy or sludgy thinking. Vanilla blocks this release and also lowers levels of neuropeptides that cause painful conditions such as arthritis, fibromyalgia, and nervous system issues.

Of course, getting the right vanilla is the issue. A lot vanilla loses its beneficial properties when destroyed by high temperatures. This also causes toxic mold particles which grow on the dried beans when stored and reduce their anti-inflammatory, cognitive-enhancing powers.

Basically, cooking vanilla takes away many of the cognitive benefits. What you want is to find a high quality, low temperature form, that reduces mycotoxins. Just be careful. The vast majority of so-called vanilla products are artificially flavored and not actually derived from vanilla fruits. For a pure and fragrant vanilla that goes great in coffee and ice cream and just about anything, try Bullet Proof's variety.

Mind Your Sides

No, this doesn't mean standing with armor and making sure nobody catches you off guard. What it does mean is practicing a little 'peripheral perception.'

Start by sitting in a location outside your place of residence. Go to a park, on a blanket or bench or in a café or restaurant. Focus on staring straight ahead, without moving the eyes. Although your eyes don't move and are facing ahead, try to see everything in your field of vision. This includes things or people moving off to the sides of your vision.

When you're done, try to write down everything you saw. As you do it again and again, try to add to that list. The reason this visual memory challenge works is because it actually triggers your brain to release more acetylcholine. Acetylcholine is a chemical associated with focus and memory and is one of the main lacking components in the brains of Alzheimer's patients.

Dark Chocolate

Many people love chocolate, but the reason isn't simply because of how it tastes. Sure, sugar makes this decadent treat all the sweeter, but that doesn't mean that we don't like it for other reasons. When we eat chocolate we end up activating systems in the brain that dispense dopamine, a critical brain chemical. This influx of dopamine in certain systems facilitates learning and memory. In addition, chocolate contains flavanols, what are anti-oxidants also found in red wine and berries. The darker the chocolate, the more powerful these effects.

Stop and Smell the Roses

This isn't just some hokey saying. Research indicates that natural scents take a direct route to the brain. Such aromas can come in oils but are also found all over our amazing planet. The reason these scents work is because many of them cross the blood-brain barrier and increase oxygen flow to the brain. Increased oxygen in the brain is linked to improved energy, mood, learning, memory, and immune functioning.

So go out amid the trees or into the ocean breeze. Smell the pines, the sand, the lavender, the flowers—you name it. Inhale natural fumes, and stay away from all that manmade smog! If you're in a concentrated urban area, it might be a good idea to look into incense and aromatherapy.

Or go on vacation! Or, at least, a trip to the countryside.

Neurobic Exercises

In case you haven't heard of them, neurobic exercises are techniques and activities designed to rewire your brain. Coined by Dr. Lawrence Katz, these brain-ticklers increase oxygen and foster improved learning, memory and problem-solving. For the most part, neurobic exercises ask us to step outside of our comfort zones and stretch our minds—like aerobics (or acrobatics) for the brain! This causes creative expansion by forcing us to participate in new activities, places and events. Neurobics feed new information to the brain, strengthen cognitive skills, create an appreciation for diversity, alter behaviors, boost mood, and encourage us to take on new challenges and opportunities.

Consider the following neurobic activities. Many of them are unconventional and some, downright odd. However, this is good. It forces your brain to make connections it might never have otherwise. Try the following:

- Walk backwards throughout the day

- Do not talk for a day, days, or weeks on end

- Blind Drawing – take items from around your home, put them in a box, and close your eyes. Remove them one by one, and use your fingers to approximate what the item may be. Then, draw the item as best you can. If you'd like, even take a stab at what color the item may be. Open your eyes after drawing several items to see how you've done.

Keep doing this on separate occasions to see if you get closer.

- Wake up early and watch the sunset
- Try strange foods

- Don't wear shoes or sox. This stimulates the mind as well as the nerve endings in the body. This builds up coordination and opens our receptiveness to stimuli. It is a good way to bridge the disjunction between our physical bodies and other physical entities
- Take a different way to work or school for a week; change it up from week to week

- Stop using facebook, myspace, or other social networks for an extended period

- Stop watching television for several days at a time

- Wear clothes inside or backwards, with noticeable textures, and mix up accessories; wear your hat on your arm, your socks as gloves, and belts as necklaces (this is probably best done when you're by yourself ;)

- Move your furniture around at home or work

- Turn all the lights off in your home and try to navigate at night (not recommended for people who are afraid of the dark)

- Quit smoking

- Spice Smelling – Try rounding up all the spices in your cupboards and putting them in a box. Then, making sure you're blindfolded, take them out one by one and write down which spice you think it is based on smell or taste. Keep doing this until they've all been lined up outside the box. Then, remove the blindfold and see how many you got right. Try this on separate occasions to see if you improve.

• Go to a public place alone, such as a restaurant, during a busy time

• Play with a ball. This might not seem like a particularly brain-teasing activity, but research shows that the build-up in coordination is immense, creating new brain connection for visual-spatial awareness

Memory- Boosters

Memory is critical, as you already know. Which is why it's good to know that there are many ways to enhance it. Still, before you employ the little hacks that can get you to an improved memory, remember the general guidelines. If you want to have a stronger more specific memory, you need to stay attentive. Researchers indicate that it takes roughly eight seconds of powerful focus to turn a piece of information into actual memory. For those who are distracted quite easily, a quiet place will do wonders.

Another way to maximize memory is to recruit all senses to the task. Relate information to colors, textures, smells, and tastes. Also try to make it a tactile experience by writing down what you really want to remember. Or, in today's modern age, type it into your smartphone's storage. It doesn't matter if you consider yourself a visual learner, if you write it and read it aloud, you'll remember better. For ultimate use of this auditory/tactile exchange, try to make a rhyme of the information.

Another obvious but oft-ignored memory tactic includes integrating information with what we already know. If you simply remember something because you have to, you might not remember it long. This is because your brain won't find it as significant. However, if you take something and do a little work to find why it matters and why it connects with your previous knowledge, you're *much* more likely to remember. For complex material, understanding ideas is better than memorizing separate details. Put the ideas into your own words and even explain them to others. Also be sure to rehearse information periodically after you learn it. Do your best to repeat information the day you learn it. Research shows that you will retain much more when you take a 24 hour period with sleep to learn it. 'Cramming' as many students call it, is actually very ineffective. Especially if you try to do it hours before an exam or test.

Okay, so now that you recognize the general tips, it's time to get more specific. One of the most useful devices for improving memory is the mnemonic. Mnemonics are devices that associate information to remember with an image, sentence, or word(s). They are nifty little tricks that allow us to store information much more efficiently and recall information with a little more zest than usual. One such mnemonic device is the image. This is basic. All you have to do is associate an image with what you want to remember. The more developed the image (vivid, colored,

3d, etc.), the more easy you will remember it. Say you want to remember the name of the city Aspen in Colorado. Although this example is a little crass, it is entirely conceivable that you could picture a pen—like a pen of pigs—but full of buttocks. That's right, quite literally an "ass pen." Or maybe you imagine a writing pen for the buttocks. Whatever floats your boat! The point, of course, is that you make your image creative and memorable. Nothing wrong with a little flare.

Another type of mnemonic is the acronym. This device is a word in which the letters represent what you want to remember. For instance, you could use the acronym "**DARE**" which could mean **D**rug **A**buse **R**esistance **E**ducation. Or, for younger kids, maybe you could come up with: "**D**rugs **A**re **R**eally **E**wwww." Acronyms can be used for anything. Just think of any professional sport: NFL, NHL, MLB, NBA, PGA, etc.

Another mnemonic device is the acrostic which takes a sentence and makes the first letter of each word significant. This is sort of like an acronym, but different. Take for instance the common acrostic used to remember the planets: **My Very Elegant Mother Just Served Us Nachos**. In this case, the first letter of each word represents the first letter of each planet name, Pluto not included. Still, acrostics can be simpler than this. If you simply want to remember, say, certain notes, you could come up with: Every Good Boy Does Fine for notes E, G, B, D, and F.

Method of loci

This is a technique that allows us to associate important information with locations. It sounds weird at first, but is actually quite effective. Simply visualize plopping items you need to remember in a familiar place or route. If you want to remember grocery items, imagine a carton of eggs at your garage door, milk and orange juice along the walkway to your front door, and then boxes of cereal at your front doorstep, and so on.

Consolidation

This is one way to take really long lists of information and consolidate them into manageable chunks. This one will come in handy when remembering a phone number. If you happen to lose your number or got that special somebody's number on a piece of paper and lost the darned paper, break the 10 digit number into three sets: 2345670012 becomes 234-567-0012. You can also use consolidation for remembering lists of towns or cities or factoids. How you organize depends on your preference. You could form chunks from smallest to biggest, chunk based on cities that start with the same letter, or that are in the same country, and so on.

You might be wondering why and how these brain hacks and neurobic exercises work. The basic point to remember is *functionality*. This refers to the ability of our brains to adapt to a variety of stimuli and demands. By practicing these habits and off-the-wall exercises, we keep our brains malleable and responsive. The best exercises are those that target our perceived weaknesses. If you don't eat well, it's time to do so. If you aren't utilizing your brain for things outside math, language, social studies, or some other particular domain—branch out! If you want to improve your focus and your reflexes, your memory and your creative thought, then apply the appropriate hacks!

And if you're simply getting older, slower, and more tired—read on! Now is *not* the time to get discouraged…

Rewind for Wisdom -- 7 Practical Strategies to Boost the Aging Brain

Not all of us are old and decrepit.

But still, why wouldn't we try to slow the effects of aging? Although researchers once believed that our brains became deteriorating sacks past a certain point, we now know that brain plasticity makes a lifetime of learning possible. Just as we can keep our bodies fit with age, we can do the same with our brains. Nothing like a good mind sweat to get the neurons crisp and loose.

So whether you're young, middle aged, or a senior citizen, it's time to feel young again!

Here are some practical strategies for boosting your aging brain and getting on the right cognitive track.

Eat Away

You've heard that food is good for the body and brain, but did you know that the right food can *significantly* enhance cognitive skills? The best food to feed your brain should be based on a natural regimen, containing good fats found in olive oil, nuts, and fish, as well as lean protein.

But it's more than that. You need to know what to gobble up and what to leave behind. If you're looking to really boost your brain power with the right diet, there are some things to keep in mind:

Cut down the calories, shave the saturated fat

Diets chock full of saturated fat increase the risk of neural deterioration linked to dementia and Alzheimer's. This means that you should limit red meat, whole milk, butter, cheese, and ice cream. Moreover, stuffing your face with too many calories will also bog down concentration and memory. Eating too many calories as you age is particularly problematic.

Drink green tea

Green tea has been touted since the very beginning for its incredible health benefits. It is a delicious and relaxing drink that stimulates the mind in a way that is calming and sustainable. Due to the wealth of polyphenols in green tea, the drink has a powerful antioxidant effect on our brain.

Green tea is also great because of the combination of caffeine and the amino acid L-theanine. Caffeine works by increasing the firing of neurotransmitters like dopamine and norepinephrine, improving vigilance, reaction time, mood and memory. Meanwhile, L-theanine works by complementing the rare negative side-effects of caffeine. In other words, L-theanine has an anti-anxiety effect, which staves off any possible jitteriness caffeine would cause in, say, a strong coffee. In the end, brain function improves and people capitalize on a mild, more stable buzz that is arguably more effective than that offered by coffee.

The Power of Omega

That's the power of omega-3s. These fatty acids are great for brain health, as supported by numerous studies and experts across the nutritional world. The best source of omega-3 is fish such as salmon, halibut, mackerel, sardines, tuna, trout, and herring. Other sources include flaxseeds, pinto beans, winter squash, broccoli, soybeans, and walnuts. There are plenty of heart healthy dishes and appetizers that include these fatty acids.

The Rainbow

That's *the rainbow* of fruits and vegetables. As you probably already know, produce is jam-packed with antioxidants, which protect brain cells against damage. The leafy green vegetables such as spinach, chard, arugula, ale, broccoli, romaine lettuce and fruit such as bananas, cantaloupe, watermelon, apricots, and mangoes are incredible for your brain.

If you want to be especially creative, you can even include these delicious fruits and vegetables into specialty drinks.

For a quick boost of brain-jolting, body-pumping goodness, experiment with fruit infusions and select juicing recipes.

Wine

That's right, wine, baby. The trick here, of course, is to limit your intake. You can't simply guzzle or the positive effects will go away. Experts recommend 1 and 2 glasses for women and men per day, respectively. This helps because the resveratrol in the grapes—especially in red wine—increases blood flow to the brain. If you don't want the alcohol effect, opt for grape juice, cranberry juice, or simply grapes and berries.

Meditate

A lot of people will look at meditation and think that it's stupid. How can relaxing for a bit really be as beneficial as they claim?

Well, the answer might surprise you. Meditation is a cornerstone of many Buddhist philosophies as well as other Easter traditions. Recently, it has begun to cross over into the Western World, popularized by the preternaturally powerful Mindfulness movement.

Meditation works for a reason. When practiced the right way, it actually changes the very structure of the brain. The brain plasticity that occurs as a result of mindfulness targets specific areas of the brain. Grey matter density and thickness actually increase in the frontal cortex and insula, reversing atrophy that occurs with age. In fact, consistent meditation has been shown to keep these brain areas as fresh as they are in people 20 years younger. Neuroimaging reveals that the capacity for decision-making, planning, and judgment (among other higher-thinking abilities) vastly improves as a result of meditation.

Brain Games

You've probably heard of Lumosity.com and other brain-boosting websites that tug and tease the very limits of your cranium. Basically, these websites teach the brain to work faster and more efficiently, thus slowing the decline associated with diminished cognitive use. Another such website that uses close-focus exercises with high rewards is postscience.com.

Basically, videogames are not mind-numbing. Despite the stigma many gamers continue to receive, there is a lot of good in some of their habits. In fact, Nintendo was inspired by the research of a Japanese doctor who wanted to develop a game called Brain Age: Train Your Brain in Minutes a Day. The game has sold millions of copies in Japan.

To keep your brain young and malleable, you can do tons of other things. You can do jigsaw puzzles, crosswords, learn a new language, start a reading habit, take music lessons, build a piece of furniture, learn martial arts, conquer a new mathematical domain, study new words, etc.

The trick is to do something that you want to do. And if you're convinced that you simply don't like to learn, you're wrong. Think harder. There is probably something you do every day that involves a degree of learning. Pick up new hobbies. Get into fantasy sports. Follow certain statistics. Simply surfing the internet has been shown to boost the mind, as the active learning associated with different websites forges new connections that not even traditional reading can.

Openness to new experiences is critical when testing your brain.

In fact, Johns Hopkins researchers have found that openness to [10 sessions of cognitive training](#) can boost an elder's reasoning ability and speed-of-processing for up to a decade following the training.

Participants in the memory training group learned ways to remember word lists and sequences of items, text material, and the story main ideas and details. Those in the reasoning group learned how to solve problems and recognize patterns; and those in the speed-of-processing group engaged a computer program that dispersed visual information at a rapid rate, forcing the mind to quickly discriminate relevant information from extraneous.

The findings indicate that these three types of training groups are instrumental in changing daily behaviors. As a result, the elders become more self-sufficient, active, and curious. This prevents social isolation, loss of fitness and fear, thus extending longevity and improving both physical and mental health. More specifically, the Johns Hopkins study indicated that memory performance improved for up to five years post-intervention, compared to 10 years for reasoning and speed-of-processing.

Other studies have shown that cognitive training programs also boost a person's willingness to experience a variety of new things. Openness to experience is one of five major personality traits, with others being: agreeableness, conscientiousness, neuroticism and extraversion. Unlike these other four, however, openness to experience has been shown to change throughout a person's life with the right interventions. This is because such openness is about being flexible and creative, taking on unfamiliar intellectual and cultural pursuits in order to explore new horizons.

Laugh

Pretty simple, right? Well, not always. Some people get so caught up in all the tedium of their daily lives, that they forget just how good they may have it. They forget to smile and laugh when life offers those wonderful little moments that might never come again.

Amazingly, laughter triggers various regions across the brain, allowing people to think and associate more freely.

And it doesn't take a comedian to introduce more laughter into your life. Simply putting a smile on your face—'forcing' one—will fill you with more positive emotions. This, in turn, will increase the likelihood that you laugh things off.

Of course, that's not to say *don't* listen to funny material. Research shows that listening to jokes and making sense of punch lines activates areas of the brain critical to learning and creativity.

To squeeze more laughter into your life, try these easy measures:

Laugh at yourself

Don't be afraid to show your embarrassing mistakes and moments. If you can point to the times you took yourself too seriously, do so. Understanding that life is to be enjoyed not dreaded, is the first step in letting loose and finding the funny in everyday happenings.

Seek the Funny

Don't avoid laughter. People will often relate their own funny stories once they've heard yours. This will ensure that you not only connect with somebody, but do in a light-hearted self-revelatory fashion that can only be good for you. Not to mention, seek out people who are always humorous. They will make fun of life's absurd little qualities and will definitely keep you in better spirits.

Never grow up

Who better to model off of than children? These innocent little creatures never seem to lose their sense of wonder. If you find yourself caught up in the drivel of adulthood, take a step back. Observe children at play, laughing, playing, and taking things less seriously.

Sure, you probably can't afford to be this way *all the time*—but you sure can some of the time.

Surround yourself with funny faces

Whether it be a clown doll, a humorous saying, a computer desktop, funny photo albums—it doesn't matter. What matters is that you are nearby something that makes you feel less stressed and more happy about your life.

Don't forget to give this special something some of your time when things get especially hectic.

Active Social Life

Looks like your friends will not only keep you happier, but healthier as well. Because humans are extremely social animals, having relationships are a key to our survival. Beyond that, however, relationships stimulate our brains. Harvard research reveals that people with the most active social lives exhibit the slowest rate of memory decline. Incredibly, said memory decline among the most integrated occurred at less than 50% the rate of those who were more isolated.

The reasons for this should come as no surprise. Being around a variety of people will cause your neurons to excite. You'll be subjected to different experiences, thoughts, beliefs, challenges and circumstances—all of this changes your neural networks. Even having a pet, like a sociable dog, has been shown to slow the rate of cognitive decline.

If you're struggling to get out there, there are many easy ways to do so. The first way is to stop watching so much T.V. Get away from your computer and television or handheld devices, and get out into the day. Meet people face-to-face, elect to communicate with your friends or acquaintances in person instead of through devices. You'll feel the *real* connection this way. Make sure to be with people who are positive and interesting, as well as interest*ed* in your wellbeing. Be sure to keep tabs on how your friends, colleagues, family members, and neighbors are doing.

Voluntarism is another great way to get your social life in order. Altruism, or the act of helping others while expecting nothing in return, is a great way to feel happier and younger. You not only expand your life, but you realize how much power there is in even the simplest things. Schools, churches, nonprofits, and charitable organizations will be more than happy to have you. Most of these people are wonderful human beings by nature and will make you feel more than welcome.

Finally, feel free to take a risk and jump head-on into some other social network. Join a political group, find a special interest group, participate in recreational sports or hobbies, find a fitting team—anything that you like that has a common interest. Don't be afraid to try something new, and don't be afraid to try something else if your first attempt isn't what you wanted.

Exercise

This one is also talked about, but it's so true that it's worth repeating. Maybe you need to simply read the literature. Boston University School of Medicine (BUSM) researchers recently discovered that certain hormones triggered by exercise directly affect the area of our brain responsible for memory and learning. More specifically, hormones tied to aerobic exercise such as running, biking, and swimming.

The study essentially measured the blood hormone levels of healthy adults. It correlated their hormone results with their level of aerobic fitness, finding that hormones called "growth factors" were released in the brain's hippocampus. Endurance exercise also increases levels of a chemical called irisin, which crosses the blood-brain barrier and activates cognition genes. In turn, this stimulates neurogenesis and brain plasticity, making for a brain that is not only healthier but more adept at new tasks.

Aerobic exercise also triggers lower levels of cortisol, the stress hormone. This is good news, as continued high levels of cortisol lead to anxiety, depression, and other neurological problems. Moreover, our neurons can even atrophy as a result of chronic stress, meaning that chronic stress may very well make us less intelligent, or at the least, less able to harness our intelligence. This is why experts contend that aerobic exercise has an impact on *fluid intelligence*, the ability to readily adapt to new situations and problems in real-time. This differs from *crystallized intelligence*, the knowledge base of facts and life experiences from which we draw.

If you've ever been so stressed that your mind is sluggish and your memory foggy, you know what this is like.

Sleep

You know that sleep is a must. You also probably know that you aren't getting as much as you should. That's why it's important to act quickly. Sleep is critical, whether a young child with a brain like a sponge, a teenager going through puberty, an adult caught in the throes of the work week, or an aging person staving off cognitive decline.

The quality of our sleep—the slow-wave activity that occurs during deep sleep—gradually decreases as we age. In neuroscientific studies at the University of California,

Berkeley, age-related changes in sleep and brain structure have been linked to reduced memory.

The study was structured so that participants, some young, some older, were tested on memorized word pairs. After testing, the participants went to sleep and their brain activity was measured. Following 8 hours of sleep, the subjects were again tested on these word pairs and measured via fMRI scans to reassess brain activity.

The result revealed that older people showed slower brain-wave activity and lesser memory capabilities than the younger group—which is not surprising. Furthermore, older adults received less benefit from sleep than did younger people, and scans revealed that they were actually using a different part of their brains for memory than younger people.

This means that optimizing sleep is even more critical for older people than previously thought. In addition to the other practical strategies discussed, sleeping well will keep the mind sharp and fresh—ready to take on new challenges and integrate them into the neurogenic brain. Fortunately, older adults who want to maximize the depth and length of their sleep can do so drug-free, assuming they've learned the right [corrective strategies](#).

Alright, so there you have it.

We've covered some of the most thoroughly researched strategies for improving the quality of your brain, strategies that can get you from feeling foggy and slow, to clear and quick. If you want to be able to keep your faculties intact as you age, you should certainly take the holistic approach. Embrace a new lifestyle as you age, opt for healthy, delicious food options. Throw away preconceptions and challenge yourself. Continue to seek new avenues for personal growth, new experiences and attitudes and behaviors. Learn as much as you can and treat your brain like you would your body. Work it out, let it recover, and never take it for granted!

But what if you want something else?

What if you're seeking a more subtle form of brain power? Perhaps you're an experienced person, but you still feel like you're lacking. Maybe you go through life, doing your best to get ahead by using your noggin, but things *still* aren't falling into place.

Maybe you need a little extra help. Not only to get your brain working better, but to put that brain to use around others. See, in today's world, success is more than just crunching numbers, using words, completing tasks and cashing checks. In today's highly interconnected and complicated world, success is directly dependent on one's ability to work with others.

In some cases, a little manipulation, or shall was say *'persuasion'* can go a loooong way…

<u>Dark Psychology</u>: Unleashing the Power of Manipulative Mind-Control

Generally speaking, manipulation is a no-no.

But this is reality we're talking about. Sometimes, things aren't so nicely black and white. In the majority of our crazy lives, everything is a sea of gray. You need to do the right things and say the right things to get ahead in today's world—whether it's in your profession, in your personal life, or just out and about, shopping for a bargain on household supplies.

If you can show up to day-to-day life armed with the right tools, it's hard to fail. Well, it's hard to fail *completely*. Sure, we'll all experience a degree of failure, no matter how successful we think we are—but that's good!

Failure will teach us to become better. It will show us how *not* to accept mediocrity and underperformance. So open your mind! Cast away preconceptions and allow yourself to get a little naughty. Sometimes we have to use our powers to sway others in our favor. Sometimes, a little dark psychology can shed light on our crazy little worlds.

And by adopting the following tactics, you will *drastically* improve your chances of capitalizing on the emotional ineptitudes of others…

Positive Persuasion

This refers to the ability to use positive reinforcement in a way that creates agreeability in your *target*. Before you begin to go to work on your coworker, opponent, friend, acquaintance, significant other, or family member—remember to positively reinforce!

In the beginning, positive reinforcement is like a warm hug. By encouraging behaviors that you want with nice words, small tangible rewards, and physical contact, you can quickly put your target in a position of vulnerability. The target will feel that he or she has your approval and will unconsciously align his or her behaviors with what you want to see. This will also put your target in a positon that does not expect the later use of dark psychology.

It is a great way to set up somebody for later persuasion, especially if that person used to think that you didn't like him or her.

False Disclosure

This refers to the tendency to reveal supposedly personal information early on in an interaction or relationship. By doing so, the manipulator gains more respect and trust from the target. Unfortunately, this information is only disclosed to gain control over the target. In most cases, the information is false or misleading, and the manipulator will withhold anymore, keeping the target tagging along.

Negative Persuasion

No, this doesn't refer to doing something bad to your target. What it actually means is to offer reinforcement when your target *stops* doing something you don't like. For instance, if your coworker is partying more often and coming to work late, improving your chance of the promotion, facilitate this behavior through encouragement. In other words, reward your coworker for his stopping of good behaviors (working hard, coming in early, sucking up to your boss). In your case, these so-called good behaviors are actually negative because they increase competition and reduce your chance of winning your boss's highest approval. Negative persuasion is also used in many marriages or relationships where one partner will offer sex acts in exchange for the other partner's ceasing of unwanted behaviors.

Varied Persuasion

This concept simply refers to the tendency to use positive and negative persuasion at unexpected intervals. This is key if you want to truly persuade your target. Why this works is pretty simple. Like a rat in a box hitting a lever for food, the more varied your reinforcement, the more likely the rat is to hit the lever. If the rat knows that food will always dispense, it will get bored. Likewise, if the rat knows that food will never dispense, it will also stop hitting the lever.

But if the rat gets food sometimes, and not others, it will continue to hit that lever at expected intervals, even though you are rewarding at *un*expected intervals. This tactic can be used on your target in a million ways. Just imagine people at casinos. The reason they keep playing—and in some cases, get addicted—is because they never know when that next big win is coming. They know it *could* come at any time, so they are pulled along, wanting and wishing almost pathetically for their big chance.

Many so-called 'unhealthy' relationships feature varied persuasion, in which the manipula*tor* keeps the manipulat*ed* strung along, creating feelings of doubt, anxiety and fear. In the end, this only furthers the extent of power and control over the target.

Muted Attack

The manipulator will approach the target cleverly by using criticism in a non-threatening manner. In other words, the manipulator will verbally attack the target but without the tone, pitch or loudness normally associated with such attempts. The effect can be very powerful. The target is more likely to internalize the insult as truth if it is uttered in a calm and even manner. For instance, if the manipulator wants to accuse the target of being irresponsible, he or she may state matter-of-factly, "You're a clueless child," instead of shouting *"You're a clueless child!"*

The manipulator may even go a step further, explaining to the target the he or she is trying to help the target. In reality, this is not the case, but the manipulator will claim the moral high seat, convincing target that this is an important moment for learning and improving.

Emotion-Shaming

This is a tactic that can be employed to discredit virtually any argument. In cases where the target seems to be losing his or her cool, emotion-shaming shifts the focus from the merits of the argument to the delivery of the argument. In other words, it doesn't matter what the target is saying. If he or she is using excessive emotion, or unwanted emotion, the manipulator can calmly point to said emotion as unmerited. Furthermore, the manipulator may express that said emotion is unfit for not only the current situation but all situations. The manipulator will often express that emotions exacerbate the problems causing such disagreements; in short, the manipulator avoids the argument altogether.

The target is made to feel overemotional, uncontrolled, and childish. The manipulator will relate all excessive displays of emotion to inferiority, even if the situation *does* merit such displays. In the end, the target may internalize his or her emotions, becoming inexpressive and confused.

Competitive Coercion

The manipulator knows that it takes more than two to play the game. That's why the manipulator will bring others into the fray. No matter what it is, whether it's a sexual relationship, a working relationship, a friendship, family relations, or a mere acquaintance, what matters is implication. That is to say, the most important factor is implicating others. If the target knows that another man or woman is of sexual interest, jealousy ensues. This causes competition and a willingness to appease the manipulator. If the target knows that other coworkers, business partners, family friends, heirs, customers, or opportunists are now contending, that target will likely do more to appeal to the manipulator.

Because the manipulator has already established a relationship with the target, the target is more likely to try to please the manipulator than directly address the lesser known competitor.

Empty Promises

These don't necessarily have to be promises, per se. They can be expressions of friendship, love, happiness, strength, or any other positive emotion. Phrases such as "I love you," "Of course I will," "You can count on me," "I promise," and "You can trust me," are ones used over and over to gain the target's trust. In almost all cases, the manipulator has become so desensitized to these words that they no longer mean anything. He or she will spout them off without a second thought, at the perfect time, making the target feel valued and connected.

Blatant Denial

Blatant denials are, well, quite blatant. Essentially, the manipulator will outright deny having said or done something, even when the target remembers it clear as day. In many cases, the target will even have other forms of evidence, and even upon directly confronting the manipulator with said evidence, the manipulator will be steadfast in denial. In the end, the target will begin to question his or her interpretation of reality, despite the overwhelming evidence.

Turning the Tables

This refers to the manipulator's tendency to turn an indictment or criticism back on the target. If the target says something along the lines of, "You've been very secretive as of late, are you hiding something?" the manipulator will immediately turn this around. Even if the target has every reason to think and/or say something, the manipulator will make him or her feel foolish. Many times, the manipulator will coolly explain that the target is being "insecure," "worrisome," "insensitive," or "overreacting" and/or "exaggerating." In rare cases, the manipulator may react with an insinuative comment that indicts the target for something unrelated or imaginary: "You accuse *me*, yet you're the one who does x, y and/or z."

Detachment

This is used when the manipulator wishes to punish the target without words. Instead of saying what he or she wants or feels, the manipulator will refuse to speak, listen, or simply withdraw physically. This causes the target to wonder what is wrong, and often leads to the target going to great lengths to 'make up' for the perceived wrong. In some cases, the target will try to appease the manipulator without having any idea of what the problem may be. Many manipulators will do this periodically, even when there is no real issue, simply to control the target.

Emotion Explosion

This is a very visceral, powerful attack that will leave the target shocked, scared, and even paralyzed. What the manipulator does is respond with a disproportionate emotional outburst to the target's actions or words. Usually, this response is expressed in a loud, angry tone, causing the target to suddenly snap quiet. It will leave the target unprepared for future explosions, and increase the likelihood that the target does not repeat the same behavior. The fear of future explosions keeps the target 'in-check' so that the manipulator maintains the upper hand.

The Body Language Retort

Even the simplest body language can significantly alter another person, and the manipulator knows this. When discouraging unwanted behavior, the manipulator will respond will nonverbal communication that resonates strongly. This may include waving the hand dismissively, wagging a finger, scoffing, rolling the eyes, delivering a pretentious or smug expression, and/or looking away in uncaring fashion. The most skilled manipulators will pair body language with verbal communication to readily influence even the most stoic targets.

The Smiling Predator

This refers to the ability to seduce or influence a person with charm. It happens in companies across the world, relationships of every type, and in daily interactions on all the streets, towns, and cities we know. The Smiling Predator uses a style that is powerful and subtle. He or she will employ hypnotic, trance-like abilities by focusing solely on the target. The target will feel as if he or she is the only thing that matters, and the predator will attach with incredible ease, giving the illusion of a real and deep connection. The ability to stare right into the eyes of the target and say whatever is needed—the right business proposition, words of understanding, feelings of love, a reassurance—*this* is what makes the predator so powerful.

At the end of the day, manipulation is about understanding how to get people to think and act the way you want. It's not necessarily about being evil or thinking evil, but it does require a degree of dark psychology. That is to say, you have to accept the idea of getting people to do things they wouldn't otherwise do. Whether or not you rationalize this as being good for them, for you, or for the both of you—that's up to you—but what matters is how you execute.

If you practice and if you start slowly, you can work your way into the psyches of many unwitting persons. It's all about building momentum and using that forward progress to sway the masses as you see fit. Of course, if your plan from the start is flawed, you could very well end up in a big bath of hot water… or a vat of acid!

But maybe you don't want to be a manipulator. Maybe you simply want to know what to look for, and if that's the case, you're in good company. By understanding the aforementioned tactics, you now have a clearer idea of what to expect and how to react. The trick is, never let a manipulator into your mind. Let the manipulator *think* he or she has the winning hand, but never show your cards. That is, until you've had enough.

Manipulate the manipulator, and get in life where you want to be!

But enough about manipulation.

This book was written to help you improve your cognitive skills, your thinking skills, and your rational approach to the emotional undercurrents that sometimes catch us all. If you apply the hacks, strategies, tips, and science discussed in this book, you *will* improve your mental and physical health as a result. Remember, the brain and body go hand in hand. How you think is how you feel. And how you feel is often tied to how you look to others—and to yourself.

Life is about taking the power that knowledge can afford you, and running with it! If you want to try surgery or medication instead, that is certainly an option. As science evolves, such options will likely become much easier and much less risky. However, if you are seeking natural means to improvement, the kind of lifestyle changes that boost the depth *and* breadth of your thinking, this book will serve you well.

A Special Note:

Thank you for reading "*Master Mind: Unleashing the Infinite Power of the Latent Brain.*" If you enjoyed reading this book and would like to be included on an email list for when similar content is available, feel free:

[Sign-Up Now]

As always, thank you for reading.

And may you continue to live healthily and happily.

Sincerely,

C.K. Murray

Other works by C.K. Murray:

1. [Health Hacks: 46 Hacks to Improve Your Mood, Boost Your Performance, and Guarantee a Longer, Healthier, More Vibrant Life](#)

2. [Body Language Explained: How to Master the Power of the Unconscious](#)

3. [Deep Sleep: 32 Proven Tips for Deeper, Longer, More Rejuvenating Sleep](#)

4. *The Blood Pressure Diet: 30 Recipes Proven for Lowering Blood Pressure, Losing Weight, and Controlling Hypertension*

5. *Coconut Oil Cooking: 30 Delicious and Easy Coconut Oil Recipes Proven to Increase Weight Loss and Improve Overall Health*

6. *High Blood Pressure Explained: Natural, Effective, Drug-Free Treatment for the "Silent Killer"*

7. *The Wonders of Water: How H2O Can Transform Your Life*

8. *INFUSION: 30 Delicious and Easy Fruit Infused Water Recipes for Weight Loss, Detox, and Vitality*

9. *The Ultimate Juice Cleanse: 25 Select Juicing Recipes to Optimize Weight Loss, Detox and Longevity*

10. *The Stress Fallacy: Why Everything You Know Is WRONG*

11. *ADHD Explained: Natural, Effective, Drug-Free Treatment For Your Child*

12. *Depression, Drugs, & the Bottomless Pit: How I found my light amid the dark*

13. *Confidence Explained: A Quick Guide to the Powerful Effects of the Confident and Open Mind*

14. *Sex Science: 21 SIZZLING Secrets That Will Transform Your Bedroom into a Sauna*

15. *Sex Secrets: How to Conquer the Power of Sexual Attraction*

16. *How to Help an Alcoholic: Coping with Alcoholism and Substance Abuse*

17. *Success Explained: How to Seize Your Moment & Take Back Life*

18. *Vitamin D Explained: The Incredible, Healing Powers of Sunlight*

19. *Last Call: Understanding and Treating the Alcoholic Brain (A Personal and Practical Guide)*

20. *Hooked: Life Lessons of an Alcoholic and Addict (How to Beat it Before it Beats YOU)*

21. _Master of the Game: A Modern Male's Guide to Sexual Conquest_

22. _Natural Weight Loss: PROVEN Strategies for Healthy Weight Loss & Accelerated Metabolism_

23. _Mindfulness Explained: The Mindful Solution to Stress, Depression, and Chronic Unhappiness_

Printed in Great Britain
by Amazon

42000022R00046